LANDFORM
No. 4
Classic Landforms of The Weald

D. A. Robinson and R. B. G. Williams
University of Sussex

Published by the Geographical Association
in conjunction with the
British Geomorphological Research Group

THE GEOGRAPHICAL ASSOCIATION
343 FULWOOD ROAD, SHEFFIELD S10 3BP

Contents

	Page
Introduction	5
The geological structure and relief of the Weald	6
The Hythe Beds of the western Weald and the Vale of Fernhurst	9
The karstic features of the Hythe Beds outcrop south of Maidstone	14
The Weald Clay lowland and Greensand escarpment south of Sevenoaks	17
Landslipping around Wadhurst	21
Sandstone cliffs of the High Weald	25
Ashdown Forest	35
The High Weald coast from Hastings to Pett	39
Glossary	47
Bibliography	47

Plates

Aerial view of Older Hill, near Redford	12
Aerial view of a shallow landslip at Little Trodgers, Mayfield	24
A nineteenth century view of the Toad Rock, Rustall Common, Tunbridge Wells	30
Sandstone cliffs at Bowles Rocks, near Eridge	32
Aerial views of the coast at Lovers' Seat taken in 1978 and in 1981	42
Lovers Seat, Fairlight, 1849	43

Editorial Introduction

Geomorphology is the study of landforms and the processes that shape them. In Britain we have many beautiful examples of classic landforms that are regularly visited both by educational parties and by the general public. The aim of this series is to provide concise, simple and informative guides to these features, so that visitors can understand them and enjoy them all the more. We hope that the diagrams, maps and glossary will make the guides useful to a wide audience, and we also aim to provide information on such matters as access, safety, conservation and the best viewpoints.

The relevant 1:25 000 maps for the sites dealt with in this Landform Guide are Ordnance Survey Sheets SU 82/92 and 83/93, TQ 33, 42/52, 43/53, 74, 75, 80 and 81. At the 1:50 000 scale, the area is covered by Sheets 186, 187, 188, 197 and 199.

Andrew Goudie, *School of Geography, University of Oxford.*
Rodney Castleden, *Roedean School, Brighton.* (Series Editors).

Other titles in the Landform Guides series:
No. 1: Classic Coastal Landforms of Dorset
by Denys Brunsden and Andrew Goudie
No. 2: Classic Landforms of the Sussex Coast
by Rodney Castleden
No. 3: Classic Glacial Landforms of Snowdonia
by Kenneth Addison

Cover photograph
The Boar's Head, a sandstone pinnacle between Eridge and Groombridge (see page 33).

The opinions expressed in this publication are those of the authors and do not necessarily represent the views of the Geographical Association.

© The Geographical Association 1984.
ISBN 900 39587 7.

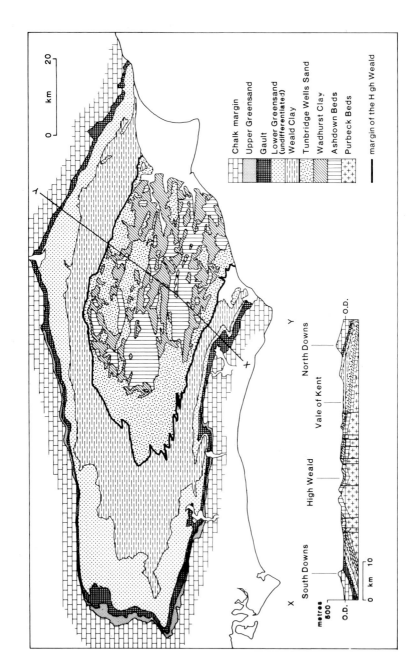

Fig. 1 The geology of the Weald.

Classic Landforms
of
The Weald

INTRODUCTION

The Weald has long been regarded as a classic area for geomorphological studies. Until recently the majority of these studies were primarily concerned with the evolution of the drainage system and the development of erosion surfaces. This approach led to S. W. Wooldridge and D. L. Linton proposing a complex model for the denudation chronology in a now classic monograph entitled 'Structure, Surface and Drainage in South-East England' (Institute of British Geographers 1939, republished by George Philip 1955). This model has recently been modified and updated by D. K. C. Jones in his book 'Southeast and Southern England' in the series entitled 'The Geomorphology of the British Isles', published by Methuen in 1981.

For many geomorphologists the focus of interest has nevertheless shifted to studies of the ways in which the processes of weathering and erosion, both past and present, have acted upon the various rock types exposed within the region to produce the distinctive landforms visible at the present day. This brief guide attempts to illustrate this second approach by describing landforms, or groups of landforms, from selected locations within the Weald, and outlining their probable origins, as far as they are known. The selection of sites for detailed description has been made on the grounds of scientific interest, accessibility and the proximity of other landform features within a small area. Landforms similar to those described in the text can frequently be found elsewhere in the Weald and wherever possible we list some of these alternative sites at the end of each section.

THE GEOLOGICAL STRUCTURE AND RELIEF OF THE WEALD

The Weald forms the heartland of south-east England. It comprises an oval area of sands, sandstones and clays encircled by the high chalk escarpments of the North and South Downs and the Chalk uplands of the 'Western Downs' of Hampshire (Fig. 1). With the exception of three small inliers of Jurassic rocks around Battle and Heathfield, all the exposed rocks belong to the Cretaceous System and range in age from 135m to 100m years (Fig. 2). All these rocks, and the overlying Chalk, were uplifted and folded into a complex anticline or dome in late Cretaceous and Tertiary times. Viewed broadly, the rocks within this dome dip very gently to the north and south, at an average of only 1–2°, away from a central axis which runs approximately east-west through the high ground of Ashdown Forest and Crowborough Beacon (TQ 512307). Locally, however, the dip of the rocks is much greater due to the presence of a number of subsidiary folds which are aligned approximately parallel to the direction of the main aniclinal axis. In addition, the rocks are often faulted along the line of many of these subsidiary folds.

Considerable erosion occurred at the end of the Cretaceous and in later times. The chalk was stripped off the Wealden anticline, and the underlying rocks exposed. The oldest rocks are found in the centre of the now breached anticline and progressively younger rocks outcrop to the north, south and west. In the east, the Weald is terminated by the present day coastline, but the geological structure crosses the floor of the English Channel and is continued in the Bas Boulonnais in north France.

Differential weathering and erosion of the various rocks exposed in the Weald has produced an area of diverse relief and scenery. The highest ground no longer follows the axis of greatest uplift in the centre of the Weald, where the maximum elevation is 240m at Crowborough Beacon, but lies some 12km to the north on the crest of the Hythe Beds escarpment of the Lower Greensand at Leith Hill (TQ 139432) which reaches 290m.

The Lower Greensand is remarkably variable in composition. The basal unit, the Atherfield Clay is generally thin, but impermeable, and produces a spring line at the base of the overlying Hythe Beds. In east Kent, the Hythe Beds consist of bands of hard limestone 'Rag' alternating with bands of soft sand and sandstone 'Hassock', and give rise to some interesting karstic phenomena. Westwards from Maidstone, however, the limestone dies out and the beds consist of sand and sandstone reinforced in many areas by bands of chert. Where the sandstone and chert are particularly well developed, the Hythe Beds build a bold and impressive escarpment which exceeds 250m in height in a number of places. In contrast, between Reigate and Dorking (TQ 165495 to TQ 277495), the beds are mainly composed of uncemented sands which offer little resistance to weathering and erosion, and give rise to a very subdued escarpment. In Sussex, the Hythe Beds

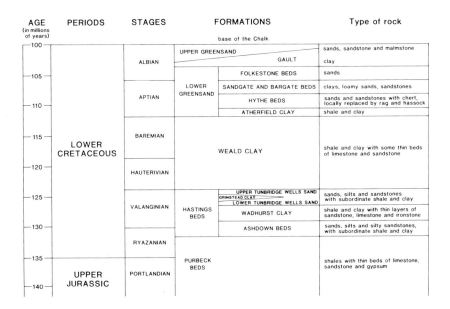

Fig. 2 The rock succession in the Weald.

thin dramatically as they are traced from west to east, and from the Arun gap eastwards (TQ 025125), they have very little effect on the relief.

The Sandgate and Bargate Beds, which overlie the Hythe Beds are generally more loamy than the Hythe Beds and tend to form relatively subdued relief. The succeeding Folkestone Beds are coarse, mostly uncemented sands which frequently form low ridges covered by heath or woodland. Beyond the outcrop of the Folkestone Beds lies a discontinuous vale underlain by the Gault Clay, followed by the Upper Greensand which sometimes forms a marked bench or low escarpment at the foot of the Chalk Downs.

The Lower Greensand escarpment faces inwards towards the centre of the Weald and overlooks a belt of lowland developed on the Weald Clay. This is the Low Weald, a rather featureless area which has received relatively little attention from geomorphologists. It forms a horseshoe-shaped surround to the central upland core, known as the High Weald, which is underlain by the Hastings Beds. The latter is an area of relatively flat-topped ridges and deeply incised valleys which cut through a complex series of alternating sands, sandstones and clays. In some locations the sandstones stand bare of any soil cover to form valley side cliffs and crags, whilst the clay slopes are frequently oversteepened and suffer from mass movement.

The relief and structure of the Weald can be observed from a number of excellent viewpoints. In the south, the best viewpoints are on the crest of the South Downs and include, from east to west, Firle Beacon (TQ 485059), Ditchling Beacon (TQ 332140), Devil's Dyke (TQ 258112) and Chantry Hill (TQ 086120), all of which have vehicle access and adequate parking for cars and minibuses. The views from each of these locations encompass the Low and High Weald, and on very clear days visibility extends to the Hythe Beds escarpment and the North Downs. In the north, extensive views over the Weald can be obtained only from the central section of the North Downs, notably from Box Hill (TQ 179513), Reigate Hill (TQ 255520) and White Hill (TQ 325536). East of Oxted (TQ 3845222) and west of Dorking, the view south from the North Downs is restricted by the high ground formed by the Lower Greensand. As a result, the Weald Clay Vale and the High Weald can only be seen from viewpoints on the crest of the Hythe Beds escarpments such as Gibbet Hill (SU 90358), Holmbury Hill (TQ 105429) or Leith Hill in the west, and Crockham Hill (TQ 448513), Toys Hill (TQ 470513), Ide Hill (TQ 486515) or Carters Hill (TQ 558531) in the east.

THE HYTHE BEDS OF THE WESTERN WEALD AND THE VALE OF FERNHURST

In the Western Weald, the sands and sandstones of the Hythe Beds are massively reinforced by lenticular bands of subordinate chert. The chert is very resistant to erosion and the Hythe Beds build bold escarpments which reach 280m at Blackdown, (SU 919296) and 272m at Gibbet Hill (SU 900358). In this region, local anticlinal flexures are superimposed on the main Wealden axis causing the outcrop of the Hythe Beds to widen from less than 2km south of Godalming to more than 12km in the vicinity of Hindhead, Haslemere and Blackdown. Along the north of the outcrop the Hythe Beds emerge from beneath a cover of Bargate Beds and rise southwards to form the high ground of Hindhead Common and Gibbet Hill which marks the crest of the Hindhead Anticline. Southwards from here, both the geological strata and the land surface dip down into the Haslemere Syncline where the base of the Hythe Beds lies some 55m lower than on the crest of the anticline. South of Haslemere the strata and the land surface rise again, ending in a high escarpment at Blackdown which overlooks the low ground of the Vale of Fernhurst (Fig. 3).

The Vale of Fernhurst provides a particularly spectacular example of anticlinal breaching. The eastern part has been opened up by erosion which has broken through the gently dipping southern limb of the asymmetric Fernhurst Anticline

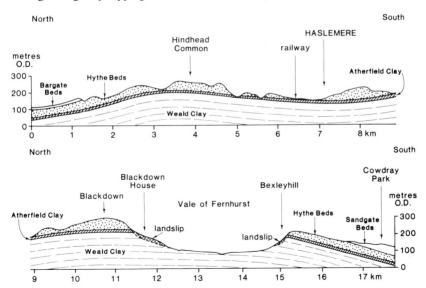

Fig. 3 Geological section across Hindhead Common and the Vale of Fernhurst. (See Fig. 4 for the line of section across the Vale of Fernhurst.)

exposing the Atherfield and Weald Clays. The axis of the anticline lies close to the crest of the escarpment at Blackdown (Fig. 4). On the south side of the Vale the Hythe Beds form a prominent north facing escarpment which is cut through by the River Lod at Lodsworth (SU 935230). The western end of the Vale has been opened up along the axis of the Harting Coombe Anticline which lies some 2 to 3km to the south of the Fernhurst Anticline. The Vale is open to the east where it joins the Weald Clay Plain but is closed in the west at Harting Coombe where the Hythe Beds have not yet been breached and the north and south facing escarpments come together.

Throughout the entire Vale, the inward-facing escarpments, which rise up to 150-200m above the floor of the Vale, are never more than 4km apart. The Hythe Beds have been *cambered* (bent) down the scarps and there are a number of landslips, notably around Blackdown in the north and around Older Hill (SU 909253) in the south. The origin of both these features remains uncertain. Cambering is believed to be a fossil feature which is no longer developing under present day conditions. It probably originated during the Quaternary when the underlying clays froze during each prolonged period of cold and then thawed when the climate ameliorated again. Upon thawing, the clays would have become wet and yielding, and probably incapable of supporting the overlying sandstone which consequently bent down the scarp face and broke open along the pre-existing joints.

The massive rotational landslips may have originated under similar conditions. However, some of these remain at least partly active at the present day with some slipping still occurring during periods of unusually wet weather and spring thaws. In addition, smaller translational landslips and mudflows often develop on the lower clay slopes of the escarpments.

Cambering is best demonstrated at Bexleyhill. The regional dip of the Hythe Beds at this location is approximately 5° to the south. Rocks dipping in this direction can be observed in an old pit 200m or so back from the crest of the scarp a little to the east of the road which crosses the scarp (SU 908251). However, blocks of Hythe Beds exposed along the sides of the road just below the crest of the scarp (SU 908253) dip down-scarp at 10-11° whilst at the crest of the scarp the interbedded sands and sandrock are horizontally disposed. The Hythe Beds thus appear to have bent down the scarp and some blocks have broken away and moved downslope.

The largest of the landslips lies along the south-west face of Blackdown. It is a massive rotational slip which extends for over 3km and affects the whole depth of the scarp between 210m and 90m O.D. The degraded scar at the back of the slip forms a steep upper slope, developed primarily in the Hythe Beds. Below this, slipped material forms a complex mass of hummocky debris which stretches southwards over the more gentle footslopes of the scarp which are developed in the underlying Atherfield and Weald Clays. The dividing line runs close to the road that skirts the scarp face of Blackdown.

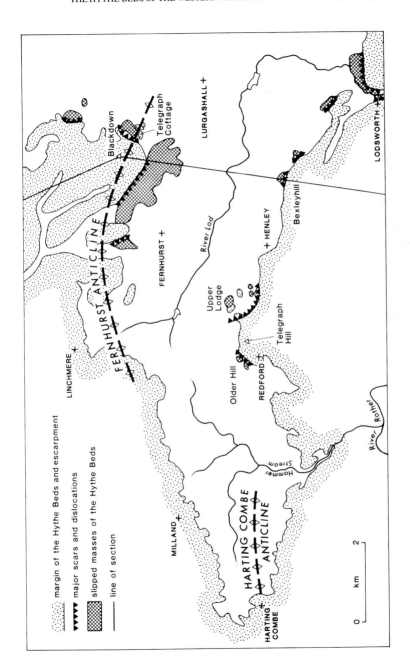

Fig. 4 Landforms of the Vale of Fernhurst.

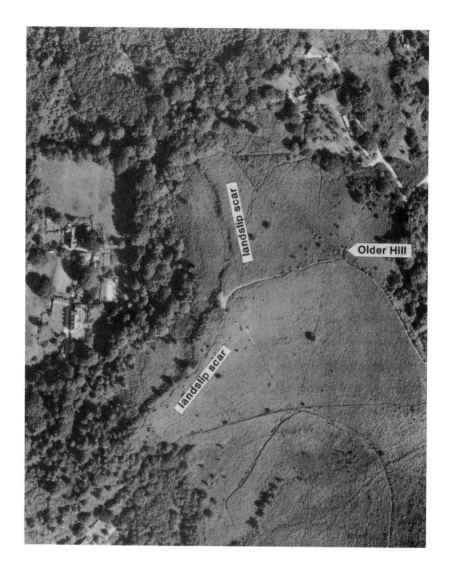

Plate 1
Aerial view of Older Hill, near Redford, showing a prominent landslip scar, probably modified by quarrying, on the western slope. (Photograph supplied by Meridian Airmaps, Lancing, reproduced by courtesy of West Sussex County Council.)

The Hythe Beds on the east side of Blackdown have also suffered major dislocation and displacement probably as a result of both cambering and land-slipping. Telegraph Cottage (SU 924291), for example, lies on a detached outlier of Hythe Beds which has been displaced downslope and tilted so that a westward dip of 45° can be seen in a roadside exposure on its southern side (SU 924290). To the north of here, subsidence of Hythe Beds along the eastern slope of Blackdown has produced a pronounced linear scar and terrace feature which runs just below the crest of the slope for more than 350m (from approximately SU 921292 to SU 921295).

On the south side of the Vale of Fernhurst slipped and cambered blocks of Hythe Beds are clearly visible on the western side of Telegraph Hill (SU 873263). The blocks descend into an unnamed, low-level valley which forms a major breach in the escarpment. Older Hill, the summit of which lies some 22m below Telegraph Hill, is the highest of a series of ancient slipped and cambered blocks of Hythe Beds which form a ridge-like promontory on the eastern slope of the valley. At the western edge of this promontory, the steep slope overlooking Redford (SU 865261) has slipped much more recently and there is a well defined, arcuate scar which exposes 4m or more of rubbly Hythe Beds (Plate 1).

The dip slopes of the Hythe Beds are dissected by a series of valleys whose upper reaches, at least, are usually dry. The road south from Bexleyhill follows an excellent example of such a valley and another can be seen on the northern slope of Blackdown. On the north slope of Hindhead Common lies the unique valley of Highcomb Bottom which terminates in the spectacular coombe-like amphitheatre of the Devil's Punchbowl (SU 895360). The stream in this valley has cut right through the Hythe Beds into the Atherfield Clay beneath. It is fed by springs at the foot of the Hythe Beds and the coombe-like head is the result of spring sapping and associated slumping of the Hythe Beds. The outcrop of the Atherfield Clay is almost completely masked by slipped and downwashed sandy debris from the Hythe Beds. The northerly gradient of the valley floor is less than the northerly dip of the Hythe Beds. Northwards, this results in the valley floor crossing onto the Hythe Beds and the whole valley narrows thereby accentuating its bowl-like head.

Access

All the sites mentioned are close to minor roads and are directly accessible by car or minibus; but access points by coach are more distant. There is ample off the road parking for cars and minibuses at the summits of Bexleyhill and Older Hill, but parking is very restricted along the road that skirts the face of Blackdown. The road to Older Hill is particularly narrow and the junction with the Redford to Woolbeding road (SU 866251) is best approached from the south. If coaches are used, Blackdown is best reached on foot from Fernhurst, Older Hill from the road at Redford, and Bexleyhill from a lay-by on the A286 at the summit of Henley Hill (SU 894255).

THE KARSTIC FEATURES OF THE HYTHE BEDS OUTCROP SOUTH OF MAIDSTONE

In the Maidstone district the Hythe Beds consist of 'Kentish Rag' or 'Ragstone', a hard-grey limestone, and 'Hassock' which is a soft, often calcareous or clayey, sand or sandstone. They form a prominent south-facing escarpment that exceeds 100m O.D. for long distances.

Karstic processes have affected the Ragstone at many places on the dip slope. Several small streams flow for short distances down dip before disappearing into tiny swallow-holes. There are other, larger depressions well away from streams that appear to have been caused by solution subsidence. The largest is Langley Hole (TQ 807522) which is about 300m long and about 10m deep with near vertical, rocky sides. It was probably formed when the roof of an underlying cavern collapsed.

The Loose Valley, south of Maidstone, is one of the most peculiar in south-east England and is well worth a day's detailed exploration (Fig. 5). It begins near the village of Langley (TQ 805516) and runs westwards towards Loose, roughly parallel with the escarpment to the south. At Loose it turns north away from the escarpment and follows a zig-zag course down dip to join the valley of the Medway.

The upper part of the Loose Valley near Langley is only shallowly incised into the dip-slope and is unexceptional in appearance. A small stream rises on the floor of the valley at a point which migrates seasonally according to the level of the water-table in the Hythe Beds. At a point north of Brishing Court (TQ 777518) the stream disappears underground down a small swallow-hole that is normally full of water. Oil put in the stream at this swallow-hole in the last century is said to have reappeared two and a half hours later at a spring some 2km away in Boughton Park, below the Hythe Beds escarpment. To reach this spring the water must flow against the dip of the Hythe Beds descending from one bedding plane to another.

A few metres down valley from the swallow-hole, water issues from a small spring and flows a short distance westwards along the valley before disappearing down another swallow-hole. The source of the water and its ultimate destination are unknown. To the west of the second swallow-hole, the valley is streamless for just over half a kilometre and has steep, generally cliffed sides. This stretch is known as Boughton Quarries, and is thought by some to have been much modified by ancient quarrying. However the name may be misleading for valleys with a similar gorge-like form can often be found in other limestone districts where they are considered to be of natural origin. The floor of the Boughton Quarries section of the valley has a reversed slope for a short distance, but whether this is because it is underlain by quarry waste or has been affected by solution subsidence is uncertain.

At the western end of Boughton Quarries (TQ 769518) there is a steep drop in the floor of the valley where a curving step crosses from one side to the other. A

KARSTIC FEATURES OF THE HYTHE BEDS OUTCROP SOUTH OF MAIDSTONE 15

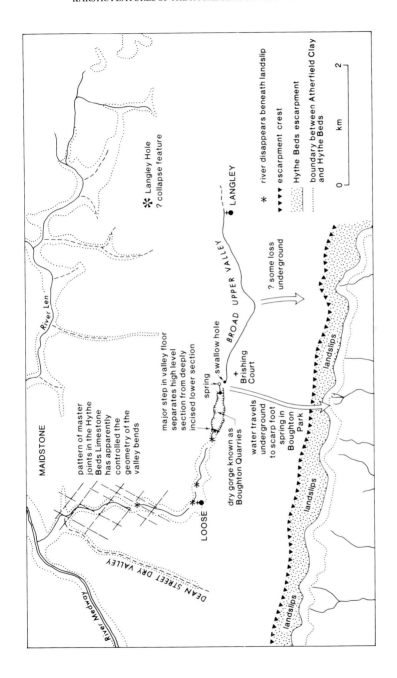

Fig. 5 The landforms of the Loose Valley south of Maidstone.

sizable stream issues from springs at the base of the step and flows off down valley to the west. The source of the spring-water is problematic. The only sizable stream that disappears underground anywhere in the vicinity is the stream flowing down the upper part of the Loose Valley, and at least some of this water appears to emerge in Boughton Park. However, some of the water may continue to follow the Loose Valley under Boughton Quarries to emerge beneath the step. An alternative possibility is that the springs at the base of the step are fed by rainwater infiltrating into the Hythe Beds at Boughton Quarries and the surrounding areas.

Below the step the Loose Valley maintains a fairly uniform gradient down to its confluence with the Medway. One explanation of the step is that it may be a *knick-point* developed by the back-cutting of the lower valley in response to a fall in base-level. As the knick-point worked headwards it lowered the water-table in the Hythe Beds further up the valley causing the stream which excavated the valley to disappear underground above Boughton Quarries. An alternative hypothesis is that the step has been caused by spring sapping at the junction of the Hythe Beds and the underlying Atherfield Clay which outcrops on the floor of the valley beneath the step. At the present day the springs at the base of the step are presumably causing a limited amount of retreat, but whether they could account for the retreat of the step up the entire valley is uncertain.

The stream that flows away down valley from the step disappears and reappears at least three more times before finally reaching the Medway. If the valley is floored by impermeable Atherfield Clay, as indicated on the Geological Survey map, it is difficult to explain these disappearances. However, they seem to occur where the stream meets ancient landslip debris on the valley floor. This debris is mostly derived from the Hythe Beds and solution may have enabled the stream to create an underground route through each landslip.

North of the village of Loose the valley has a pronounced zig-zag course which is probably inherited from an earlier period when the stream was still cutting down through the Hythe Beds and had not yet reached the Atherfield Clay. Master joints in the Hythe Beds, trending both NE-SW and NW-SE, seem to have controlled the direction of flow of the water and the position of the valley bends. Later, cambering caused the Hythe Beds to bend down the valley sides.

Access

A public footpath passes the swallow-hole north of Brishing Court and joins a minor road that runs along the valley floor through Boughton Quarries. Parking is limited on this road. The section of the valley between the Quarries and Loose is not well served by paths but a number climb the valley sides and give views of the stream and the valley floor. Below Loose a footpath follows the stream to Hale Place. Langley Hole lies on private farmland.

THE WEALD CLAY LOWLAND AND GREENSAND ESCARPMENT SOUTH OF SEVENOAKS

The B2176 at Bidborough north of Tunbridge Wells (TQ 565435) runs along the crest of a high ridge formed by the Hastings Beds, which descends steeply on the northern side of the River Medway. The road provides a series of good viewpoints over the Medway valley and the low-lying outcrop of the Weald Clay on the far side. Some shallow valleys can be seen on the clay outcrop which drains southwards into the Medway. Between the valleys the ground rises slightly, forming lows hills and ridges. In the far distance, the Hythe Beds of the Lower Greensand give rise to an impressive escarpment, which maintains a crest line at around 200m O.D.

At Penshurst (TQ 528435) the River Eden joins the River Medway. The upper course of the River Eden is excavated into the Weald Clay, as is the course of the Medway east of Tonbridge (TQ 590465) but in the vicinity of Penshurst the two rivers cross the axis of the Penshurst Anticline and cut into the Hastings Beds, forming a number of fine, incised meanders. The discordant drainage is sometimes thought to have been initiated when the Weald Clay outcropped at a higher level and occupied the crest of the anticline. The two rivers can be visualised as meandering freely across the clay outcrop, uninfluenced by the anticlinal structure. After the rivers had succeeded in crossing the fold, base level fell and they began to incise their courses. The Weald Clay outcrop was greatly lowered, and where the rivers crossed the fold axis the underlying Hastings Beds were exposed.

North-east of Penshurst stands Camp Hill (TQ 423468) one of the low hills on the Weald Clay outcrop. It is capped by a layer of clayey cherty debris up to 1.5m thick. No good sections are available at present, but numerous angular fragments of chert can be picked up on the surface of ploughed fields. The chert is derived from the Hythe Beds on the escarpment some 5km to the north, and has been transported to its present location by *solifluction* (Fig. 6a).

Many other hills and and ridge tops on the Weald Clay outcrop have similar cappings, and it is thought that the solifluctional debris once extended as a continuous sheet from the base of the escarpment across the clay outcrop for distances of up to 6km on slopes inclined as low as 1°. The subsequent incision of streams on the outcrop has dissected the sheet, leaving cappings on the hills and ridges between the stream valleys. The solifluction is believed to have occurred during the last but one glacial stage, the Wolstonian, while the valley incision is thought to date from the last, Ipswichian, interglacial.

Spreads of clayey cherty debris are found on the floors of the valleys that dissect the clay outcrops at distances up to 1km from the escarpment. Chert fragments can be seen, for example, in the sides of the small stream near the public footpath just east of the village of Sevenoaks Weald (TQ 535513). These low-lying spreads of debris must postdate the incision of the valleys, and were probably soliflucted off the escarpment during the last glacial stage, the Devensian.

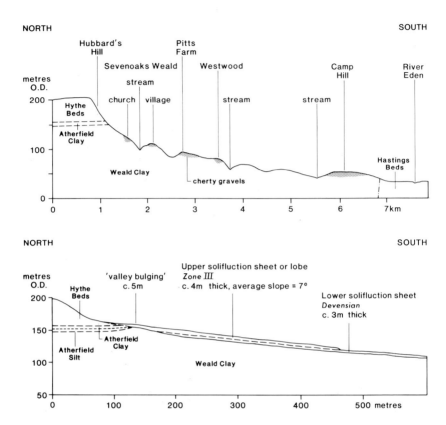

Fig. 6a and 6b Cross-sections of the Hythe Beds escarpment and the Weald Clay vale in the vicinity of Sevenoaks, Weald, between Sevenoaks and Tonbridge. Upper section, Hubbards Hill to Camp Hill; lower section, detail of solifluction lobe between Beechmont Bank. (Redrawn, with minor modifications, after Skempton and Weeks, 1976.)

A series of prominent solifluction lobes (Figs. 6b & 7) can be seen on the lower slopes of the Greensand escarpment between Hubbards Hill (TQ 527522) and River Hill (TQ 541521). They are composed of clayey, cherty material that was sludged downhill during the final cold sub-stage of the Devensian (Zone III of the Late Glacial). The frontal portions of the lobes overran and buried a soil layer

THE WEALD CLAY LOWLANDS AND GREENSAND ESCARPMENT SOUTH OF SEVENOAKS

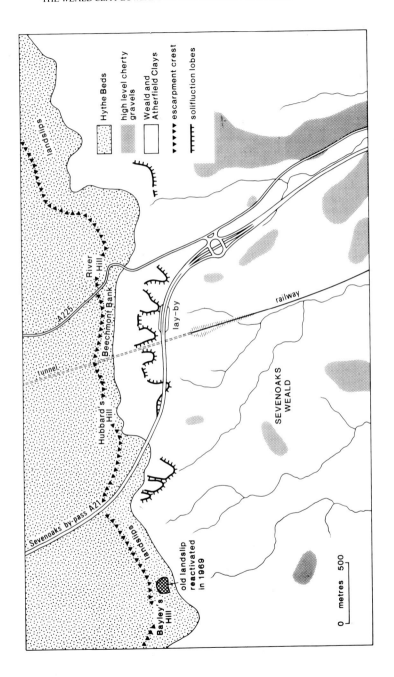

Fig. 7 The Hythe Beds escarpment and the Weald Clay vale in the vicinity of Sevenoaks Weald.

which was developed on top of the main Devensian solifluction. Humus within the soil has been dated by radiocarbon to 12,250 years B.P. (Zone II of the Late Glacial). Pollen analysis of the soil suggests the presence of light birch woodland with grassy clearings or a grassland with birch copses. The fact that a soil was able to form at this stage of the Devensian suggests that solifluction had largely or entirely ceased. When cold conditions returned solifluction resumed, and the lobes began to descend the slopes. The rapid passing of the Devensian after this date meant that the solifluction soon came to an end.

The lobes at Hubbard's Hill were discovered during the construction of the Sevenoaks by-pass in the mid 1960s which was originally planned to follow a route across the head of the lobes. Major slipping started to occur, however, and a new route had to be selected lower down the slope where it would pass below most of the lobes. The original route can be traced across the hillside by stretches of modern fencing which contrast with the otherwise continuous hedgerows around the fields.

The Hythe Beds above the lobes in the embayment of the escarpment at Beechmont Bank (Fig. 7) have undergone large-scale landslipping. The generally wooded nature of the ground, and thick undergrowth, make investigation difficult, but the slips appear to have been modified by solifluction during the Devensian and may therefore be ancient features. Subsurface investigations in the spur of the escarpment at Hubbard's Hill have shown that the Hythe Beds there have subsided vertically along joints or fractures into the underlying Atherfield and Weald Clays, which have been squeezed out and bulged. These large-scale structural disturbances probably formed in the Wolstonian and appear to be the result of the freezing and thawing of permafrost.

Landslips are still occurring on the face of the Greensand escarpment in many places. An old slip at Bayley's Hill (TQ 515518), for example, became reactivated in 1969, causing damage to roads and property. Similar slips have occurred in the recent past at Ide Hill (TQ 490513), Toys Hill (TQ 468509) and other locations. An old rotational slip forms a small back-tilted knoll on the face of the escarpment at Hanging Bank (TQ 500514).

Access

The lobes at Hubbard's Hill are on private farmland, but one of the best examples can be seen from the lay-by on the north side of the dual carriageway, south of River Hill (TQ 537517). A short walk westwards along the foot of the road embankment brings another large lobe into view. Both lobes have steep edges which show signs of recent slight instability.

Landslipping is widespread on the Hythe Beds escarpment throughout Kent, Surrey and West Sussex. Good examples not previously referred to in this guide include the south faces of Crockham Hill (TQ 446511), Tilburstow Hill (TQ 346501) and Leith Hill.

LANDSLIPPING AROUND WADHURST

In the High Weald, many text-book examples of shallow rotational and translational landslips can be found on the Wadhurst Clay (Fig. 8). The slips generally develop in winter during periods of heavy rainfall, and during thaws following prolonged frost. They are commonest at the junction with the overlying Tunbridge Wells Sand where seepage occurs and slopes are particularly steep and ill-drained. When recently formed, they tend to have prominent, bare, vertical scars at their head and pass downhill into hummocky or ridged ground which ends in a steep bank. If no further movement occurs the slips become progressively more subdued in form, and after only a decade or so they are often impossible to identify.

Roads crossing steep wet hillsides on the Wadhurst Clay have often been damaged by landslipping. Several have had to be temporarily closed or reduced to one-way traffic to allow repairs to be made. At Best Beech Hill, near Wadhurst, (TQ 619317), a minor road crossing a steep clay hillside so frequently suffered minor landslipping and required so many repairs that in about 1970 the highway authorities decided to close it permanently to traffic. Fortunately, an alternative road existed lower down the hillside which was less severely affected by slipping and could be kept open for traffic. The abandoned road has become rather overgrown with vegetation in recent years but is still interesting to visit. No slipping is visible at the top of the hill where the road is cut in the Lower Tunbridge Wells Sand. Damage is found only where the road crosses onto the Wadhurst Clay. The hillside at this point has a mean slope of about 14 or 15°, and an irregular hummocky profile due to the presence of flow lobes (Fig. 9). The road surface is broken by gaping cracks, and many of the intervening strips of road have subsided downhill, often with backward rotation. The back tilt of some of the tree stumps in the copse above the road confirms that rotational slumping is occurring. The damage to the road dies away northwards as the hillside flattens out.

The lower road is crossed by many long cracks, especially where it passes over a small stream (TQ 617316). The cracks are regularly infilled with tar, and the sides of the road have been partially rebuilt.

A good example of a translational landslip can be seen in pastures at Little Trodgers, Mayfield (TQ 589295). The slip developed in particularly wet winter weather in the late 1960s when the surface layers of the clay were thoroughly saturated. Long cracks in the pastures once marked the head of the slip (Plate 2) but these are now infilled and grassed over. However, the steep bank marking the toe of the slip still remains an impressive feature.

Another shallow slip has affected the eastward facing slope of Green Hill (TQ 575294). A fence running down the slope has been broken by the movement, and another fence running across the slope lower down the hillside has been strikingly offset.

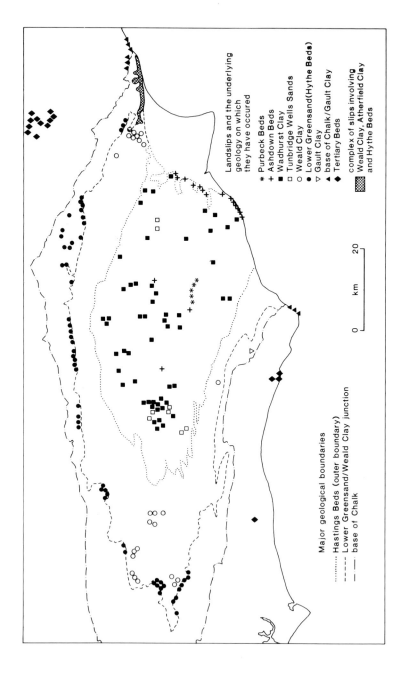

Fig. 8 The distribution of landslips and mudflows in the Weald and surrounding areas of south-east England.

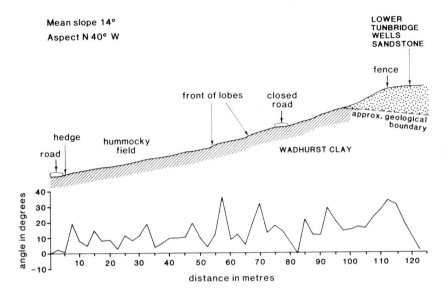

Fig. 9 Cross-section of slipped ground at Best Beech Hill, Wadhurst.

At present, landslipping on the Wadhurst Clay is confined to slopes of 12° or more, but old slip surfaces have been found up to 2 or 3m below ground level on slopes of as little as 4°. These old slip surfaces appear to be periglacial in origin, and are thought to have been created when the surface layers of clay were soliflucted downslope.

It is perhaps surprising that so many clay slopes on the Wadhurst Clay remain steep, despite the continuing landslipping and mass movement. One might have expected the slopes to have become greatly reduced in angle. It is important to note, however, that the streams and rivers of the Central Weald are rapidly incising their valleys, which helps to maintain the steepness of the side slopes. Perhaps even more important is the fact that many clay slopes are capped with Tunbridge Wells Sand which is fairly resistant to erosion and helps to maintain steep slopes on the clay beneath.

Access

At Best Beech Hill cars can be parked just to the north of the Inn on the crest of the hill, where there is an interesting exposure of flaggy Tunbridge Wells Sand. The Green Hill and Little Trodgers slips are on private farmland and permission to enter needs to be obtained from the owners.

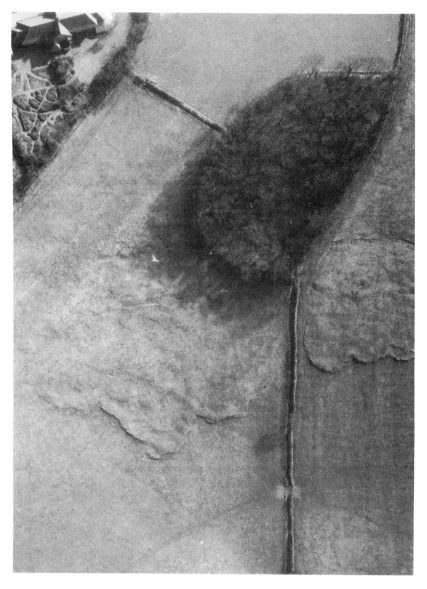

Plate 2
Aerial view of a shallow landslip in a pasture at Little Trodgers, near Mayfield. (Photograph supplied by Meridian Airmaps, Lancing.)

SANDSTONE CLIFFS OF THE HIGH WEALD

The valleys round the margins of the High Weald are incised through the alternating clays, sands and sandstones of the Hastings Beds. In some places the sandstones form bare, rounded cliffs up to 15m in height which extend along valley's sides for distances of up to half a kilometre or more. Most of the cliffs are developed in a particularly massively jointed bed within the Lower Tunbridge Wells Sand known as the Ardingly Sandstone, and their distribution, therefore, closely follows the outcrop of this sandstone (Fig. 10). However, the sandstone frequently forms steep, soil-covered slopes rather than cliffs and some cliffs are developed in hard sandstone bands within the Ashdown Beds and the Wadhurst Clay.

The cliffs have developed as a result of the relative strength and resistance to erosion of the sandstones compared to the sands and clays that lie above and below. Opinion is nonetheless divided over the mode of formation of the cliffs; they may have been formed by frost weathering and solifluction under periglacial conditions, or by weathering and mass wasting under temperate conditions. Present day processes appear to cause granular disintegration and rounding of the sandstone but retreat of cliff faces is obviously slow. Inscriptions dating from the seventeenth century remain clearly legible at a number of locations.

The tallest cliffs are found at High Rocks near Tunbridge Wells (TQ 558383) where they have been a renowned tourist attraction for close on two centuries (Fig. 11). These cliffs, which reach 15m in height, outcrop in woodland high up the slopes of a west facing spur that lies between two minor tributaries of the River Medway. Apart from their greater height, the cliffs at this location are typical of many in the Weald with bare, rounded upper surfaces and vertical sides which are frequently undercut towards the base. The continuity of the cliff faces is broken by widened vertical joints or *gulls* which separate the sandstone into a series of isolated blocks. Rock surfaces on either side of the gulls complement one another with the convexities matching concavities. This suggests that weathering and erosion have played little part in the widening of the gulls which is primarily the result of the blocks having moved apart at the joints. Movement of the blocks and widening of the gulls is believed to be due to bending and cambering of the sandstone down the slope. The joints have opened up both at right angles to the cliff face, and parallel to it, producing a labyrinth of passageways up to a metre or more in width, through which it is possible to walk. Traced back into the hillside behind the cliffs the gulls become narrower and are infilled with clay and sandstone debris.

As at many other locations, the cliffs at High Rocks are strikingly undercut towards their base. This is best displayed towards the northern end of the outcrop. Several mechanisms have been proposed to explain the undercutting, including abrasion by wind-borne sand grains, differential destruction of silty bands of rock

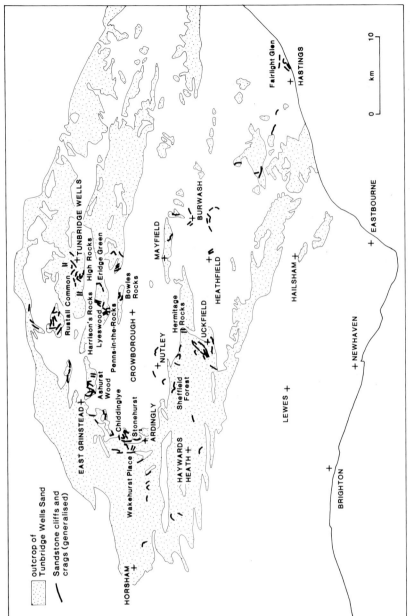

Fig. 10 The distribution of major outcrops of sandstone cliffs and crags in the Weald. Sea cliffs and abandoned sea cliffs are not shown.

at the base of the cliffs, differential rotting of the basal sandstone by water seepage, sub-surface weathering and footslope lowering, and subsurface spalling due to salt weathering by soil solutions drawn up through the sandstone and evaporated at the surface. It is a stimulating exercise to look for evidence for and against these suggested mechanisms, none of which has been conclusively proven to be the cause of the undercutting.

The cliff faces are predominantly reddish brown to black in colour due to the presence of a hardened crust or rind. The crust is enriched with organic matter from surface vegetation and with iron oxides and silica left behind when water from within the rock evaporates at the surface. It is usually only a few millimetres thick but occasionally exceeds one centimetre. In a few places the crust is absent, exposing the whitish or buff yellow underlying sandstone, which is noticeably softer and more friable. This softness is vividly illustrated by the ease with which tree branches and climbing ropes have abraded and grooved the tops of the cliffs. The crust is important to the survival of the rock because it makes the surface less porous and more resistant to abrasion.

The hardened crust has probably been forming for thousands of years and is continuing to form at the present day as is clearly evidenced by the way inscriptions carved into the face of the cliffs in the last century are now all crusted over. However, once punctured, large sections of crust can be easily destroyed because the underlying layers of sandstone are frequently weakened through the loss of material that has been redeposited in the crust. There are a number of locations at High Rocks where the crust is peeling off revealing the weakened sandstone beneath. Once the weakened layers are removed and sound rock is exposed, crusting tends to develop anew.

Where the crust is unusually thick it probably originated not as a weathering rind but as a lining on a joint face, and may be very ancient. An example of a very hard crust of this type can be seen on a tilted boulder that stands a few metres in front of the main cliff line towards the south end of the outcrop.

The vertical and undercut faces of the cliffs display a variety of cavernous micro-weathering features. The most striking is honeycomb weathering in which the whole surface of the sandstone is pitted by masses of circular hollows from one to several centimetres across. It is best developed towards the southern end of the outcrop. The majority of hollows are semi-circular in cross section, but some which penetrate the hardest crust, widen inwards to form bell-shaped hollows. In some places trails of loose sand, derived from the weakened sandstone beneath the crust extend down the cliff face.

Honeycomb weathering is usually considered to result from salt action. At first sight, the valleys of the High Weald appear an unlikely environment in which to encounter salt weathering, However, salts could be derived from inside the sandstone and concentrated at the surface by the evaporation of migrating solutions, and salt is carried inland from the coast in rain. Honeycombing is most

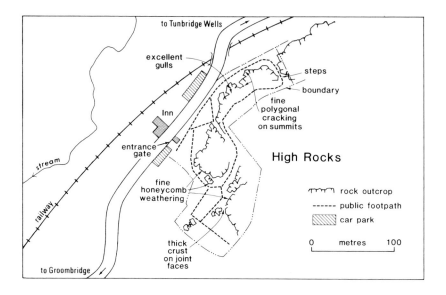

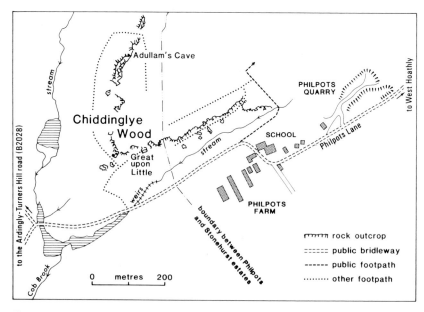

Fig. 11 Sketch-maps of the sandstone cliff outcrops at High Rocks (upper) and Chiddinglye Wood (lower).

common on sections of cliff that undergo frequent wetting and drying cycles which would favour salt weathering processes. It is absent from permanently damp rock, and cliff tops that are exposed to the sun and dry rapidly after rain.

Where bedding planes intersect the cliff face seepage of water frequently leads to the decay and disintegration of the sandstone. A succession of small cavities often develops; the cavities frequently coalesce to produce a continuous cleft, the outer edge of which may be crossed at intervals by miniature pillars or peg-like projections of sandstone reminiscent of teeth or rows of cup handles. Where several clefts are developed on a single face, the cliff begins to look like a series of gigantic woolsacks piled one upon another.

In places the faces and upper surfaces of the cliffs are dissected by polygonal network of cracks, frequently referred to as tortoise-shell weathering. On the flat and gently rounded upper surfaces the polygons are mostly hexagonal in plan but they become more rectangular on the vertical faces. The origin of the cracks remains uncertain, but it has been suggested that they might be formed by repeated expansion and contraction of the rock as a result of heating and cooling, wetting and drying or freezing and thawing. Alternatively, they could be due to changes in the volume of either the core or the casing of the sandstone as a result of weathering and crust formation.

An outcrop of sandstone at Rustall Common (TQ 568394), 1.5km north of High Rocks and on the very edge of the built-up area of Tunbridge Wells, provides some interesting contrasts with High Rocks. The site is famous for an isolated sandstone block known as the Toad Rock which stands immediately in front of the main outcrop. The block is so undercut at the base and grotesquely moulded by weathering and erosion that it resembles a large toad squatting on a pedestal (Plate 3).

The upper surface of the main sandstone outcrop is bare of any soil or vegetation cover. In places it is relatively flat and forms a sandstone pavement crisscrossed by narrow, gaping joints that have opened only a few centimetres. The sandstone is unusually impermeable, and differential weathering of the upper surface of the pavements has produced a number of weathering pits up to 0.5m across and 0.3m deep which fill with water in wet weather but dry out between rains. Cavernous weathering is poorly developed at this site, perhaps because of the low permeability of the sandstone; polygonal cracking is visible on the back of the Toad Rock.

Sandstone is also exposed in an overgrown, wooded quarry to the south. Quarrying has left vertical faces which are intensively used by climbers. The susceptibility of the sandstone to abrasion is vividly displayed by deep grooves cut by the rubbing of the climbers' ropes. On the south side of the A264, there are further outcrops of cliffs in Happy Valley (TQ 654392) where honeycomb weathering is well developed. On Tunbridge Wells Common a low upstanding mass of sandstone forms a tor-like outcrop overlooking the cricket ground (TQ 578393).

Plate 3
A Nineteenth Century view of the Toad Rock, Rustall Common, Tunbridge Wells.

In the vicinity of West Hoathly there is an extensive outcrop of cliffs and a small working quarry, known as Philpots Quarry (TQ 354321) where freshly exposed Ardingly Sandstone can be examined and compared with the crust-covered sandstone that forms cliffs in the neighbouring Chiddinglye Wood (TQ 347322). However, access is difficult and the rocks are all on private land (Fig. 11).

The fresh sandstone exposed in Philpots Quarry is buff-yellow in colour, although some joint faces are coated with white calcite and some bedding planes are coloured deep brown due to coatings of iron oxides. Many of the joints are entirely closed and none gape by more than 10cm. The rock is relatively soft, and sand grains can be detached by gentle rubbing with the fingers.

In Chiddinglye Wood, the cliffs of sandstone outcrop on the slopes of a south-west facing spur lying between two incised head-water streams of the Cob Brook 1.5km south-west of West Hoathly. The cliffs are not quite as tall as those at High Rocks, reaching a maximum of 12m. but exhibit all the same morphological features. Cavernous weathering forms are well developed, particularly honeycomb weathering. Towards the end of the spur the joint-bounded blocks of sandstone have separated to an extraordinary degree. The sandstone appears to be cambered into both valleys and down the spur producing extremely wide gulls and a chaotic arrangement of isolated joint blocks. Large boulders of sandstone are scattered on the footslopes beneath the cliffs, lying partly embedded in a talus of sandstone fragments and loose sand. Some of the boulders show fine polygonal cracking. On the north-western side of the outcrop widened joints and basal undercutting have combined to produce a cave which in former times was inhabited and is known as Adullam's Cave.

An isolated block of sandstone estimated to weigh 400 to 500 tonnes stands on a narrow pedestal a short distance in front of the cliffs on the southern side of the spur. It is separated from the main cliff face by open gulls and the pedestal is the result of basal undercutting on all sides. It is known variously as Great-upon-Little or Big-on-Little and over the centuries has attracted more interest than any other Wealden landform. It is marked on Richard Budger's map of Sussex dated 1724, and was the subject of a paper by Thomas Pownall in 1782 which is one of the first studies of any British landform. Many visitors to Great-upon-Little over the past 300 years have carved their initials, with dates, in its surface and the preservation of these initials suggests that very little erosion is occurring under present conditions.

Ardingly Sandstone is also exposed in a quarry at Hook (TQ 356314) 1km south-east of Philpots. The joints in this quarry have all opened up as a result of cambering and now gape by 30cm or more. Those exposed at the present day are infilled with clay and sandstone debris that has fallen in from above, but open voids have been encountered during quarrying. The cambered sheet of sandstone can be traced southwards, outcropping in the sides of Hook Lane as it descends into the valley of the Cob Brook.

Plate 4 Sandstone cliffs at Bowles Rocks, near Eridge.

Access

High Rocks are private but public access is permitted upon payment of a small fee at the entrance gate opposite the High Rocks Inn. The site is open throughout the year and detailed inspection of the cliffs is facilitated by a series of footpaths which give access not only to the base of the cliffs but into the gulls and up onto the summits where the gulls are crossed by bridges. Access roads are narrow but passable for coaches. Roadside parking is available at the entrance.

There is free public access onto Rustall and Tunbridge Wells Commons from the adjacent roads. Roadside parking is available but the area around the Toad Rock can become busy and congested at weekends.

Chiddinglye Wood is **strictly** private. The majority of the south-east facing cliffs belong to Mr. L. Hannah, the north-west facing cliffs and the whole of the south end of the spur, including Great-upon-Little, are in the grounds of Stonehurst, Ardingly, which is owned by Mr. Strauss. Philpots and Hook Quarries are both owned by Mr. Hannah. To date, access has normally been granted for **bona fide** educational visits on weekdays but **prior permission** must always be obtained from the owners **in writing** before entering either the quarries or Chiddinglye Wood. Mr. Hannah can be reached at Philpots Quarry, West Hoathly. Stonehurst is also occasionally open to the general public at weekends under the National Gardens Scheme. Access to Chiddinglye Wood and Philpots Quarry is from a bridle way that runs from West Hoathly (TQ 363325) to the Ardingly Road (TQ 345319). There is limited roadside parking at both ends of the bridle way but vehicles **must not** be taken down Philpots Lane which is a private road. The road leading to Hook Quarry is very narrow and impassable for motor coaches. Roadside parking is very restricted.

Both High Rocks and Chiddinglye Wood are *Sites of Special Scientific Interest.* No hammers may be taken into either site, no damage of any nature inflicted on the rock surfaces, and no moss or plants removed. No hammers are allowed into Philpots or Hook Quarries.

There are several other cliffs with public access, including:

Harrisons Rocks, Groombridge (TQ 532355)
A fine outcrop of tall cliffs with free public access. Intensive use by climbers has damaged the surface of the cliffs in places but many features of interest remain.

Bowles Rocks (TQ 545332)
An outcrop of very tall cliffs, parts of which are impressively undercut towards the base, located within the grounds of an Outdoor Pursuits Centre (Plate 4). Payment has to be made to visit the cliffs. There are further isolated outcrops a short distance to the south, including a spectacular pillar of rock surmounted by a rounded boulder known as the Boarshead (TQ 536328) which lies within the grounds of a private house but can be viewed from the adjacent road (see cover).

Stones Farm, Saint Hill (TQ 382348)
A stretch of cliff and isolated sandstone buttresses overlooking the western end of Weirwood Reservoir. A popular climbing venue with congested parking on a difficult corner.

Wakehurst Place, Ardingly (TQ 340315)
This property is owned by the National Trust and includes an extensive run of relatively low cliffs in Bloomers Valley (TQ 335317). The gardens are open throughout the year but there is a high admission charge.

Sheffield Forest (TQ 418262)
A series of sandstone buttresses with good honeycomb weathering outcrop on the valley sides close to public footpaths.

ASHDOWN FOREST

Writing in 1830, William Cobbett declared Ashdown Forest to be "verily the most villainously ugly spot I ever saw in England". Nowadays different standards prevail, and Ashdown Forest is held to be one of the most charming and unspoilt parts of south-east England. Its open heathlands and tracts of woodland attract great numbers of visitors, especially at summer weekends.

The Forest is also an excellent area for geomorphological and pedological fieldwork. Many different slope processes can be examined, including rilling, gullying, spring sapping and landslipping. The streams on the valley floors are rapidly incising their beds, and have produced many interesting examples of waterfalls, braids and meanders.

The Forest lies astride the main watershed of the central Weald. The northern slopes are drained by tributaries of the River Medway while the southern slopes are mainly drained by the River Ouse and its tributaries. One valley on the southern side of the Forest, however, has a very anomalous course (Fig. 12). It runs down past the Isle of Thorns (TQ 320305) before turning through a sharp right-angle bend north-west of Nutley (TQ 442287) and striking north and north-eastwards to join the Medway. It has been suggested that the valley above the bend was initiated by a headwater of the Ouse which has since been captured by a Medway tributary cutting headward along the route of the valley below the bend. A slight col at 135m on the ridge between Nutley and Duddleswell may mark the original course of the headwater as it made its way southeastwards. The present stream that flows from Nutley south-east towards Maresfield is interpreted as the beheaded lower portion of this captured headwater. After this suggested capture took place the valleys have been incised by at least 50m.

The high ground of Ashdown Forest is underlain by the relatively resistant Ashdown Beds, consisting of up to 230m of sands, silts, and clays interbedded with harder sandstones and siltstones. They occupy the core of the east-west trending Crowborough Anticline, which forms the centre of the Wealden Dome in this part of Sussex.

Hill slopes on Ashdown Forest are long and mostly gentle, but where the more resistant beds of sandstone reach the surface they tend to create benched slopes with steep fronts. The long hillside leading down from Gills Lap to Newbridge is interrupted by three benches apparently caused by sandstone beds. A sandstone bed running along the hillside east of Nutley Windmill forms a prominent bench, while another forms a conspicuous feature on a hillside south-west of Newbridge. Nowhere on the Forest is the sandstone resistant enough to form bare crags, but some small examples are found on hillsides just to the east of the Forest around Jarvis Brook.

Many of the soils on Ashdown Forest have a low infiltration rate, in part due to the widespread occurrence of compacted subsoil horizons. As a result, the soils

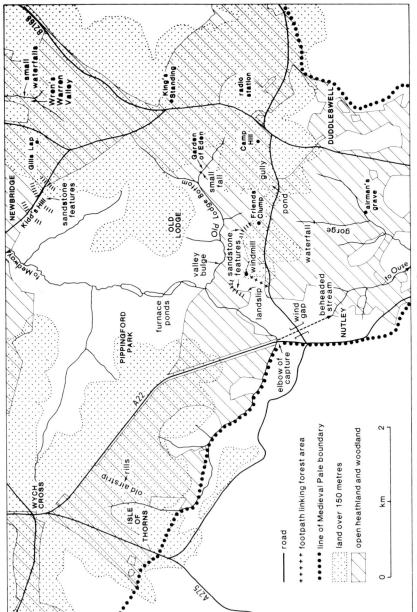

Fig. 12 Landforms of part of Ashdown Forest, between Nutley and Wych Cross.

easily become waterlogged, especially in winter when the rains are generally heavier and evaporation rates are low. Considerable overland flow can occur, leading to rilling and gully erosion, especially along paths where trampling has destroyed the surface vegetation and has increased the compaction of the soil. The best developed rills are found near the Isle of Thorns (TQ 425307) on an unsurfaced runway constructed near the end of the Second World War and almost immediately abandoned. A network of rills has cut into the runway forming miniature incised meanders. At the south end of the runway the ground steepens and the rills unite into a single channel which flows southwards through thick vegetation towards a break in slope where the channel is incised to form a gully approaching 2m deep. Waste from the gully has been deposited as a small fan on the gentle hillslope beneath the break of slope. Mapping of the rill network in 1970 showed that it had developed a channel density approaching 300km per sq km in contrast to the surrounding heathland which has a valley density of only about 3km per sq km.

Another example of a well-developed gully, this time reaching about 1m in depth, can be observed on the south side of the footpath that runs from Camp Hill Clump to the Pond (TQ 564289).

Most rills and gullies on Ashdown Forest develop by a combination of downwearing of the channels and headward erosion of miniature knick-points and steps in the channel floors. Bare ground around the channels is subject to alternate swelling and shrinkage as a result of freezing and thawing and wetting and drying, producing a broken surface which is quickly eroded by raindrop impact and sheetwash during storms.

Clay bands within the Ashdown Beds frequently produce springs or more diffuse seepage zones aligned along the contours of slopes. Sapping around the springs often produces slight hollows on the slopes, especially towards their base. At the rear of some of the seepage zones, small irregular seepage terraces with vertical faces up to 30cm in height are formed by a similar process.

Landslipping is uncommon on the Forest, presumably because there are few thick bands of clay. However, an impressive slide can be seen in a field near Nutley Windmill in an area where the Ashdown Beds are unusually clayey. This slip has been present since at least 1947 when it can be clearly identified on air-photographs. It appears to be a reactivated central portion of a rather larger, older slip, and currently displays a bare, fresh-looking head scar and pronounced frontal lobes.

Most of the permanent streams that drain Ashdown Forest are cut into bedrock and have very uneven long profiles. Even thin sandstone or siltstone bands give rise to miniature waterfalls or rapids while the softer beds are associated with more graded reaches. An unusually large waterfall occurs near the head of the valley west of Duddleswell (TQ 464280). A small stream flows down the valley floor over a series of southward dipping bedding planes developed in the top of a bed of sandstone before suddenly dropping into a well-marked plunge pool excavated in underlying sands and silts. Below the waterfall the stream runs in a miniature gorge with the sandstone exposed along the sides. The gorge has been formed by the headward retreat of the fall, perhaps in response to rejuvenation. Tributary streams also make waterfalls as they enter the gorge. There is another waterfall at the Garden of Eden (TQ 464297) and the stream in Wren's Warren Valley descends by a whole series of small falls.

The valley floors tends to be fairly free of solifluctional or colluvial debris. It is not clear whether the hillslopes and summit areas have been subject to major periglacial erosion. If they have, the debris must have been transported outside the area of the Forest, possibly by post-glacial erosion. Valley bulging, which may be produced by the melting of permafrost, seems to be only rarely developed, perhaps because most of the sediments are not sufficiently rich in clay. However, a good example of a bulge was encountered during the construction of Weir Wood dam and a tightly folded anticlinal structure resembling a bulge is visible in the bank of the stream along Old Lodge Bottom (TQ 453295).

Access

All the above sites, except for the landslip, lie on common land with free access. Ample parking is available in designated car parks scattered around the Forest. The area is easily and quickly explored because it is relatively compact and is served by a dense network of footpaths. The landslip near Nutley Windmill is in a private pasture, but can be seen from an adjacent track and adjoining common-land.

THE HIGH WEALD COAST FROM HASTINGS TO PETT

Magnificent cliffs have developed where the Hastings Beds of the High Weald meet the sea between Hastings and Pett Level (TQ 828100 to TQ 888132). Because the sea is cutting back into high ground, which reaches 145m O.D. adjacent to Fairlight Coastguard Station (TQ 862113), the cliffs are impressively tall with some vertical faces in excess of 60m in height. The form of the cliffs along this 8km stretch of coast depends upon lithological variations within the Hastings Beds and their degree of exposure to wave attack. None of the rocks is particularly hard and the average rate of cliff retreat is in excess of 1m per year. It is this rapid erosion which maintains the impressive cliffs, not the strength of the geological materials.

The cliffs are developed in the Ashdown Beds and the overlying Wadhurst Clay (Fig. 13). The lower Ashdown Beds are predominantly clays (formerly called the Fairlight Clay) with thin bands of sandstone, whilst the upper beds are predominantly sands and sandstones. The Wadhurst Clay mostly consists of shales but just above the base of the formation contains a band of massive sandstone some 10m thick, known as the Cliff End Sandstone. All these rocks lie in a faulted anticline. The axis of the anticline lies to the east of the Firehills (TQ 867114) where the oldest rocks, the clay of the lower Ashdown Beds, outcrop. South-westwards and north-eastwards from this point the cliffs are developed increasingly in the sand and sandstones of the upper Ashdown Beds and the Cliff End Sandstone.

The soft clays of the lower Ashdown Beds are rarely capable of supporting vertical cliffs. Where these beds outcrop, the cliffs tend to consist of a jumbled chaos of unstable ground stretching inland for 2-300m from the landward limit of wave action. Erosion of material from the base of these cliffs triggers successive collapses which eat back into the clay slopes of the high ground behind. The rate of erosion at the base of the cliffs is so rapid that the materials brought down by repeated landslips and persistent mudflow activity are rapidly removed by the sea and instability is maintained.

To the west of Fairlight Coastguard Station the sandstones of the top Ashdown Beds and the Cliff End Sandstone become prominent in the upper parts of the cliffs. The cliffs have a relatively low-angled footslope, developed in the clay beds, surmounted by vertical cliff faces developed in the sandstones above. The clays in the lower part of the cliff become saturated by rainfall, sea spray, and ground water which has passed through the overlying sandstones. Mudflows transport the clays forward to the sea edge where they are eroded by wave action. The loss of clay support brings down the overlying sandstones which collapse in massive rock falls. Large blocks of sandstone litter the clay footslopes and are left behind on the beach when the softer clays are washed away by the waves. Large rotational landslips occasionally occur which affect both the overlying sandstones and the clays beneath.

40 CLASSIC LANDFORMS OF THE WEALD

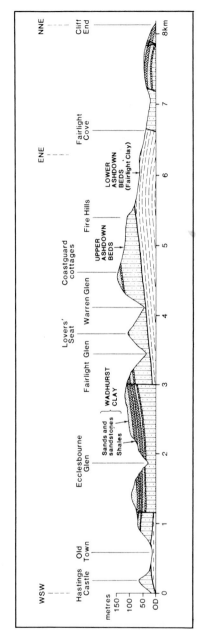

Fig. 13 Geology of the cliffs between Hastings and Pett. (Redrawn, with minor modifications, after Geological Survey of England and Wales, 1:25,000 Sheet, Hastings – Rye, Inst. Geol. Sciences, 1977.)

All these features can be observed in the vicinity of Fairlight Glen (TQ 852105) which is the only point along the cliffs with safe access to the beach (Fig. 14, Plate 5). The active mudflow zone has to be crossed to reach the beach, and this provides excellent opportunities for observing the characteristics of the flows. Their active nature is clearly demonstrated by the dislocated remnants of more than one set of access steps. The edges of individual flows are frequently delimited by sharply defined tears, side-shears, and their surfaces are often broken by transverse crevasses where they descend over convex sections of slope. Sometimes there are rafts of soil and vegetation up to a cubic metre or more in size, which lie on the surface of the flows and which have clearly been carried *en bloc* over the saturated mud below. The flows are most active in wet winter weather and least active during summer droughts. Because a proportion of their water supply comes from ground water in the sandstones above, they very rarely dry out completely, even during droughts, and are quickly reactivated when wet weather returns.

To the east of Fairlight Glen is a fresh vertical scar more than 30m in height which was created by a major landslip and rockfall during the winter of 1980-81. This spectacular cliff collapse was responsible for completely destroying a famous sandstone crag at Lovers' Seat (TQ 854108) which formerly stood close to the cliff edge (Plate 6). Mudflows at the toe of the slip project out to sea and are covered with a chaos of massive sandstone boulders. The seaward margins of the whole

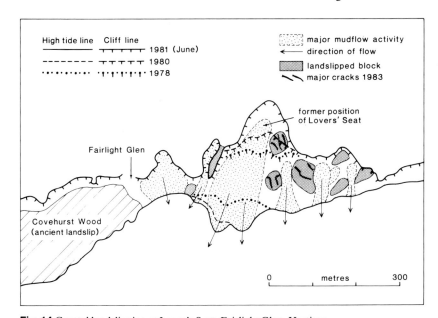

Fig. 14 Coastal landslipping at Lover's Seat, Fairlight Glen, Hastings.

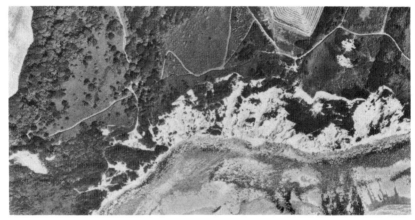

Plate 5
Aerial views of the coast at Lovers' Seat taken in 1978 (above) and 1981 (below). The spectacular extension of the mudflow and landslip which destroyed this local landmark can be clearly seen. (Photographs supplied by Meridian Airmaps, Lancing and reproduced by courtesy of the Southern Water Authority.)

interfluve between Fairlight Glen and Warren Glen to the east remain very unstable, and there are many arcuate scars 1-3m in height where sections of ground are breaking away and slipping towards the sea.

To the west of Fairlight Glen lies the massive rotational landslip of Covehurst Wood which forms an undercliff backed by nearly vertical sandstone cliffs. The landslip affects nearly a kilometre of coast and is 300m wide at its broadest point.

Plate 6
Lovers' Seat, Fairlight, 1849. This picturesque and well-known landmark was destroyed by landslipping during the winter of 1980-81. (Reproduced by courtesy of Worthing Museum.)

The slip is evidently of considerable age. It appears on a number of early nineteenth century prints by Rowe and Nicholson and on a map published by Rowe[1]. The prints show that it was clothed in dense woodland so it must date from the eighteenth century or earlier. The first ordnance survey map of 1813 shows what is presumably the slip but it is misplaced eastwards and it is not until 1868 onwards that the ordnance maps show the slip in its correct position. Low cliffs occur around the toe of the landslip where it is being eroded by the sea, and small slumps frequently occur around the seaward edge. When originally formed, the toe of the slip must have extended much further out to sea than it does at the present day.

Between Ecclesbourne Glen (TQ 837099) and Hastings, the westward dipping sandstones of the upper Ashdown Beds descend almost to sea level (Fig. 13). These massive sandstones, and the overlying Cliff End Sandstone, give rise to tall cliffs with vertical or near vertical faces up to 60m in height. Shales of the Wadhurst Clay outcrop at the top of the cliff and slump over the face of the cliffs onto the beach below.

Cliffs of a similar form, consisting of a lower vertical face developed in massive sandstones with an unstable upper zone of clays and shales, occur betwen Fairlight Cove (TQ 881120) and Cliff End (TQ 888132). Here, however, the vertical face is not as tall and is mostly developed in the Cliff End Sandstone.

An abandoned cliff line continues east of Cliff End behind Pett Level and Romney Marsh. Where this ancient cliff was developed in massive sandstone some vertical faces survive although the lower sections of the cliff are now hidden behind fallen debris. Where it was developed in clays and shales the cliff has been degraded to gentler slope angles by landslips, mud and earthflows (Fig. 8). These processes continue intermittently at the present day on all slopes greater than 10-12°. Although the clay cliffs are now entirely grassed over, except for temporary scars created by the mass movements, they form a marked feature rising above the low alluvial flats of Pett Level. The Toot Rock (TQ 893138) which stands in Pett Level, is a former sandstone islet impressively cliffed on its seaward side.

The Coastal Glens

The coastal upland between Hastings and Pett is dissected by three deeply incised valleys, Ecclesbourne Glen, Fairlight Glen and Warren Glen. Although the streams in these glens are actively lowering the floors of their valleys they have not been able to keep pace with the falling base level caused by the retreat of the cliffs. As a result, the floors of the three valleys hang some 25 to 45m above the sea and the streams issuing from the valleys cascade down the cliffs to the beach below.

The cross-sectional profile of each of the valleys is very closely related to the underlying geology. Where the massive sandstone beds outcrop on the valley

---

[1]Examples can be viewed in the library of the Sussex Archaeological Society, Barbican House, Lewes.

THE HIGH WEALD COAST FROM HASTINGS TO PETT　　　45

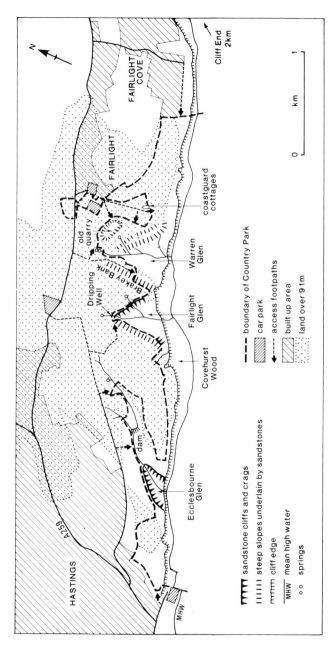

Fig. 15 The landforms of Hastings Country Park, Fairlight.

sides, they form steep slopes or lines of bare crags. Where the softer clays and shales outcrop, the slopes are gentler and often scarred by minor mass movement features.

In Ecclesbourne Glen, for example, the upper valley, excavated in the shales of the Wadhurst Clay, is broad and open with relatively gentle slopes. In contrast, the lower valley, which is largely excavated in the Cliff End and upper Ashdown Sandstones, is narrow and deeply incised with very steep slopes. The Cliff End Sandstone outcrops as a persistent line of crags which are particularly prominent on the western slopes where they reach a height of 4-6m (Fig. 15).

Fairlight Glen is developed almost entirely within the Ashdown Beds. The overlying Cliff End Sandstone forms a rim of very steeply sloping ground and outcrops as crags just below the crest of the eastern slopes. Near the head of the valley two further bands of massive sandstones are exposed beneath the crags and give rise to a double waterfall in the stream bed, often misleadingly known as the Dripping Well. Headward retreat of the waterfalls has left a narrow, gorge-like inner valley cut into the floor of the main valley.

Warren Glen is a broader, more open valley with less woodland. The slopes are developed in the alternating sandstones and clays of the Ashdown Beds, and the slopes developed on the clays frequently exhibit signs of instability. The Cliff End Sandstone forms the steep slope of Brakey Bank high on the western slope and is exposed in old quarries on the north east flank of the valley. Other bands of sandstone within the Ashdown Beds form marked benches on the lower valley sides.

Access
Most of the area lies in the Hastings Country Park and is criss-crossed by a dense network of public paths and trails. Major access points, with good parking facilities, are to be found at Hastings and Fairlight (TQ 860118). The beach is accessible from Hastings in the west and Cliff End in the east, but Fairlight Glen is the only safe access or exit along the 8km in between. Walking along the beach from Fairlight to either Hastings or Cliff End is slow and exhausting because of a scatter of large sandstone boulders and is not recommended. Good views of the cliffs can be obtained from the beaches at Hastings, Fairlight Glen and Pett, and from the coastal path along the top of the cliffs. The beach at Fairlight is used by naturists.

Safety
The cliffs are unstable and no attempt should be made to explore seaward of the well marked coastal paths and view points. Rock falls are common and it is unwise to approach close to the base of the cliffs.

GLOSSARY

B.P. The date in years before present.

Bulge. Layers of sedimentary rock bent or folded up under the floor of a valley or at the base of an escarpment. The structure is probaby produced after cold glacial phases when deeply frozen ground thaws.

Camber. Layers of sedimentary rock bent down hillsides towards valley floors. Cambering is often associated with bulging and probably also results from the thawing of frozen ground.

Escarpment. A steep slope, often 30° or more, marking the upper edge of a gently dipping layer of hard rock.

Gulls. Gaping joints in a rock that has been cambered. The joints may be open or infilled from above with sediment.

Knick Point. A reach of a stream or river that is relatively steep, resulting from a fall in base level and rapid downcutting. Major knick points are marked by falls or rapids.

Landslip. A mass of rock that has slid downhill either partially rotating (rotational slip) or moving as a flat slab (translational slip). The surface on which movement occurs is called a **slip plane.**

Mud Flow. A mass of wet clay or mud that has flowed relatively rapidly downhill.

Periglacial. An area or climate subject to intense frost action.

Rotational Slip. See Landslip.

Solifluction. The downslope movement of soil or weathered rock. The term is used most often in cold regions to describe the flow of wet thawed soil over frozen ground.

Swallow Hole. A fissure or joint in limestone that has been enlarged by solution allowing a stream to disappear underground.

Translational Slip. See Landslip.

BIBLIOGRAPHY

Bennett, F. J. (1908) Solution-subsidence valleys and swallow-holes within the Hythe Beds area of West Malling and Maidstone, *Geographical Journal*, 32, 277-288.

Gallois, R. W. (1965, 4th ed.) *The Wealden District*, British Regional Geology, Geological Survey, London. (A summary account of the geology).

Gibbons, W. (1981) *The Weald*, Unwin. (A field guide to the geology).

Jones, D. K. C. (1980) *Southeast and Southern England*, Methuen (A comprehensive account of the geomorphological evolution of the Weald and surrounding regions).

Mortimore, R. N. (1983) The geology of Sussex, in *Sussex: Environment Landscape and Society*, The Geographical Editorial Committee, University of Sussex, Alan Sutton, Gloucester, pp. 15-32. (A recent summary of the geology).

Robinson, D. A. (1971) Aspects of the geomorphology of the central Weald, in Williams R. B. G. (ed), *Guide to Sussex Excursion*, Institute of British Geographers, pp. 51-60. (An excursion route through the central Weald with a discussion of some of the major landform features).

Robinson, D. A. and Williams R. B. G. (1981) Sandstone cliffs on the High Weald Landscape, *Geographical Magazine*, 53 pp. 587-92. (An illustrated discussion of the geomorphology of the sandstone cliffs).

Skempton, A. W. and Weeks A. G. (1976) The Quaternary history of the Lower Greensand escarpment and Weald Clay vale near Sevenoaks, Kent, *Philosophical Transactions Royal Society London*, A 283, pp. 493-525. (A detailed but very readable account of the Sevenoaks by-pass site).

Williams, R. B. G. and Robinson, D. A. (1983) The landforms of Sussex, in *Sussex Environment, Landscape and Society*, op. cit. pp. 33-49 (A recent discussion of past and present geomorphological processes and the development of landforms in a major part of the Wealden Region).

Wooldridge, S. W. and Goldring, F. (1953) *The Weald*, The New Naturalist, Collins (A good but rather dated introduction to the region)

Wooldridge, S. W. and Linton, D. L. (1939, 1955) *Structure, Surface and Drainage in South-East England*, Institute of British Geographers Pub. No. 10. Reprinted by George Philip, London. (The classic model of the denudation chronology of the region, although many details are no longer acceptable to the majority of geomorphologists).

Locations of many additional sites of geomorphological interest in the Weald can be found in the following recently published Memoirs of the Institute of Geological Sciences.

Bristow, C. R. and Bazley, R. A. et al (1972), *Geology of the Country around Royal Tunbridge Wells*.

Dines, H. G. et al (1969), *Geology of the Country around Sevenoaks and Tonbridge*

Thurrell, R. G., Worssam, B. C. and Edmonds, E. A., et al (1968) *Geology of the Country around Haslemere*.

Worssam, B. C. et al (1963) *Geology of the Country around Maidstone*.